THE SCIENCE OF **SUPERPOWERS**

THE SCIENCE OF
CONTROLLING ELECTRICITY AND WEATHER

Kaitlin Scirri

Cavendish
Square

New York

Dedication: To my mother, who taught me how much fun reading could be.

Published in 2019 by Cavendish Square Publishing, LLC
243 5th Avenue, Suite 136, New York, NY 10016

First Edition

Website: cavendishsq.com

This publication represents the opinions and views of the author based on his or her personal experience, knowledge, and research. The information in this book serves as a general guide only. The author and publisher have used their best efforts in preparing this book and disclaim liability rising directly or indirectly from the use and application of this book.

All websites were available and accurate when this book was sent to press.

Library of Congress Cataloging-in-Publication Data

Names: Scirri, Kaitlin, author.
Title: The science of controlling electricity and weather / Kaitlin Scirri.
Description: New York : Cavendish Square, 2019. | Series:
The science of superpowers | Includes index.
Identifiers: LCCN 2017054205 (print) | LCCN 2017055905 (ebook) | ISBN 9781502637987
(ebook) | ISBN 9781502637963 (library bound) | ISBN 9781502637970 (pbk.)
Subjects: LCSH: Electric power--Juvenile literature. | Lightning--
Juvenile literature. | Weather control--Juvenile literature.
Classification: LCC TK148 (ebook) | LCC TK148 .S46 2019 (print) | DDC 621.31--dc23
LC record available at https://lccn.loc.gov/2017054205

Editorial Director: David McNamara
Editor: Kristen Susienka
Copy Editor: Rebecca Rohan
Associate Art Director: Amy Greenan
Designer: Joe Parenteau
Production Coordinator: Karol Szymczuk
Photo Research: J8 Media

The photographs in this book are used by permission and through the courtesy of: Cover Alexey Stiop/Alamy Stock Photo; Background (clouds) Keith Pomakis/Wikimedia Commons/File:Cumulus Clouds Over Jamaica.jpg/CC BY-SA 2.5; p. 4 Universal Images Group/Getty Images; p. 6 ©Walt Disney Pictures/courtesy Everett Collection; p. 7 Archive Photos/Getty Images; p. 10 Bettmann/ Getty Images; p. 12 John Parrot/Stocktrek Images/Getty Images; p. 14 Clari Massimiliano/ Shutterstock.com; p. 17 Encyclopaedia Britannica/UIG Via Getty Images; p. 18 Olesia Bilkei/ Shutterstock.com; p. 21 Leemage/Universal Images Group/Getty Images; p. 22 Richard Bouhet/ AFP/Getty Images; p.24 Mark Newman/Lonely Planet Images/Getty Images; p. 27 ©iStockphoto. com/AndamanSE; p 28 Ssuaphotos/Shutterstock.com; p. 29 Divanov/Shutterstock.com; p. 31 Zimiri/ Shutterstock.com; p. 34 Aerovista Luchtfotografie/Shutterstock.com; p. 37 Vitalii Nesterchuk/ Shutterstock.com; p. 39 ©iStockphoto.com/FotoVoyager; p. 40 VStock/Thinkstock; p. 42 Kckate16/Shutterstock.com.

Printed in the United States of America

CONTENTS

CHAPTER 1 HISTORY AND STORIES OF ELECTRICITY AND WEATHER 5

CHAPTER 2 SCIENTIFICALLY SPEAKING 15

CHAPTER 3 ANIMALS AND INVENTIONS 25

CHAPTER 4 INNOVATIONS OF THE FUTURE 35

GLOSSARY 43

FIND OUT MORE 45

INDEX 47

ABOUT THE AUTHOR 48

CHAPTER 1

HISTORY AND STORIES OF ELECTRICITY AND WEATHER

People have been writing stories about controlling electricity and the weather for a long time. There are stories from ancient Greece filled with gods who had superpowers. One of those gods was Zeus. Zeus had the power to control the weather. He

Opposite: In ancient Greece, Zeus was a god with the power to control the weather. This is a statue of Zeus.

could make it rain and thunder, and he could send lightning and wind to Earth. He used a thunderbolt as a weapon.

STORYTELLING

Controlling the weather is still a popular idea for stories. It can be seen in Disney films like *The Little Mermaid* and *Frozen*. In *The Little Mermaid*, Ursula, the sea witch, has the power to make a giant storm at sea. She creates lightning, thunder, and very large waves. In *Frozen*, Princess Elsa has the power to freeze things and make snow and ice. The ability to control the weather has often been seen as a superpower or a magical power.

Early experiments with electricity also gave writers ideas for stories.

Princess Elsa makes snow and ice in the Disney film *Frozen*.

One writer was Mary Shelley. She wrote a book called *Frankenstein.* In the book, a scientist named Dr. Frankenstein creates a monster. He uses electricity to bring the monster to life.

CULTURAL INFLUENCES

Trying to control the weather has been a part of some cultures for thousands of years. Some Native American cultures performed rain dances.

Ancient cultures performed a rain dance during long periods without any rain. One example is the Hopi Native Americans, shown here.

A rain dance was a special dance that would be performed during long times without any rain. Most of their food came from crops that needed rain to grow. The Native Americans hoped the rain dance would bring rain for their land and crops.

Some ancient Egyptian cultures also performed rain dances when they needed rain. They usually involved dancing in a circle. Sometimes the dancers would spin around to act like wind. Like Native American cultures, the Egyptians would perform the rain dance during dry times when they most needed rain.

DISCOVERING ELECTRICITY

Electricity was not created or invented. It was discovered thousands of years ago. The ancient Greeks discovered static electricity around 600 BCE. When they rubbed certain materials together, they noticed a spark or a shock. This was static electricity. Sometimes you can feel static electricity. This is what happens when you get a shock while touching something like a doorknob or a sweater. Or sometimes your hair might stand up or stick out.

Several people played important roles in discovering and learning about electricity. One such person was Benjamin Franklin. He is sometimes called a Founding Father because he played an important role in founding, or setting up, the United States of America. He was also a scientist and an inventor.

Benjamin Franklin studied thunderstorms and lightning. He thought lightning was a form of electricity. In 1752, he performed an experiment to test his idea. He knew that metal attracted lightning. So, he flew a kite with a metal key attached to it during a thunderstorm. He was hoping the metal key would bring the lightning close enough for him to study it. It worked, and his experiment proved that lightning was electrical.

CONTROLLING ELECTRICITY

After Benjamin Franklin's experiment with lightning, many scientists continued to study electricity. They learned new ways to use and

This illustration shows Benjamin Franklin using a kite and a key to perform an experiment with lightning.

control it. In 1800, Italian scientist Alessandro Volta created an electric **battery**. A battery is a container that holds a flow of electricity inside it. Volta was the first person to find a way to use a steady flow of electricity. This was an important invention that led to the creation of batteries as we know them today. Today we use batteries to power things such as TV remotes, cell phones, and cars.

Around 1878, American scientist Thomas Edison and British scientist Joseph Swan each invented a light bulb. These were the first bulbs that would light up for a very long time. When these scientists learned how to use electricity to create light, it changed the world. People no longer had to stop working when it got dark outside. They could work, read, and move around safely in the dark.

Edison also invented direct **current**. A current means something flowing from one place to another. An example is a river. In a river, water flows from one end of the river to the other end. Direct means going one way without changing direction. Direct current is a flow of electricity in one direction. Today, a cell phone is an example of a direct current. The electricity flows into a cell phone battery in only one direction.

Hydro Electricity

Nikola Tesla designed one of the world's first hydroelectric power plants. "Hydro" means water. "Hydroelectric" means using the power of fast-moving water to create electricity. Tesla used the powerful waterfalls of Niagara Falls to turn on a big engine. When the engine was turned on, it created electricity for nearby areas. On November 16, 1896, Tesla first used his power plant at Niagara Falls to bring power to the city of Buffalo, New York. The power plant was important because it helped electricity to travel. Before Tesla's power plant, it was hard to get electricity to people in cities far away from a source of power.

Nikola Tesla, circa 1890

In the 1880s, a man named Nikola Tesla invented a different kind of current. The current was called alternating current. When something is alternating, it means that it is changing. Alternating current is a flow of electricity that changes direction. This means that you don't get the same amount of power all the time. An example of alternating current is the electricity that gives power to your home. This is because a high amount of electricity can be sent to your home and then changed into a lower amount of electricity that is safe to be used. These early electrical inventions were steps toward people learning to control and use electricity.

CHAPTER 2

SCIENTIFICALLY SPEAKING

Electricity is a kind of **energy**. Energy means being able to move around or move an object. An example is picking up an object and moving it to a different place. It can also mean moving yourself around by walking, running, jumping, and playing. When you eat breakfast in the morning, your food gives your body energy to move around and do work at school.

Opposite: Lightning happens when electrons in a cloud and protons on the ground try to meet each other.

HOW DOES ELECTRICITY WORK?

Electricity is made up of **atoms**. Atoms are very tiny units made up of even smaller units. The smaller units are called protons, electrons, and neutrons. Protons have a positive charge. Electrons have a negative charge. Neutrons are neutral, meaning they are not positive or negative. There are two kinds of electricity. Current electricity is what happens when electrons move from one atom to another. It is sometimes called an electric current, or a flow of electricity. A second kind of electricity is called static electricity. When you experience static electricity, you feel a shock or see a spark. Usually atoms have the same number of electrons and protons. Static electricity happens when there is a different number of protons and electrons. Lightning is an example of a giant static electric

DID YOU KNOW?

The first electric **motor** was invented by British scientist Michael Faraday in 1821. A motor is a moving tool that creates power for a machine.

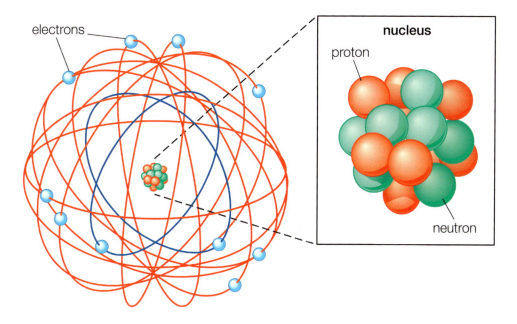

This image explains an atom. An atom is made up of protons, electrons, and neutrons. Electricity is made up of atoms.

spark. It happens when there is an uneven number of electrons and protons between a cloud and the ground. The cloud is filled with a negative charge. The ground has a positive charge. Lightning happens when these two charges try to meet each other. A giant spark, or a streak of lightning, comes out of the cloud and up from the ground.

USING ELECTRICITY TODAY

Electricity is something we use each day. When there is a thunderstorm or a snowstorm, sometimes the power will go out. This is often when we realize how much electricity we use every day. Electricity is used to turn on lights so you can see. It keeps your refrigerator running at home so your food stays cold and does not spoil. Without

Electricity keeps the food in your refrigerator cool.

electricity, cell phones and laptop computers could not be recharged and would run out of power. If you do your schoolwork or homework on a computer, then you use electricity. When you watch TV after school, you use electricity. Electricity continues to be used in new ways as well. For example, some new cars run on electricity instead of gasoline.

Sometimes, things we use need battery power to work. For example, a cell phone does not need to be plugged into an electrical source to work all the time. It needs electricity to power the battery. Once the battery is filled up with electrical power, the phone will work without being plugged in.

Not all batteries are rechargeable like cell phone batteries. Some batteries run out of power and then no longer work. Then they need to be replaced with new batteries. These kinds of batteries are often found in flashlights, clocks, and TV remotes.

Electricity is dangerous for people. It is impossible for a human to hold electricity or lightning without harm. Today, people can only control electricity through careful use of machines and batteries.

WHAT IS WEATHER?

When you think of weather, you probably think of clouds, rain, or sunshine. Weather means what it is like outside at a certain time. The weather can change quickly or stay the same for a long time. It might be sunny when you leave for school and then rainy on your way home. Or it may be cloudy and windy for days at a time.

A common type of weather is **precipitation**. Precipitation is when water falls from clouds in the form of rain, snow, or **hail**. Hail is frozen rain that falls during thunderstorms and often before a tornado. Hail can be small or large. Large hail can cause a lot of damage to crops and property and can injure people.

To understand what causes precipitation, we need to look at the science behind it. When the sun is out, it heats up water found in puddles, rivers, and lakes. When the water gets warm, it rises up into the sky. This process is called **evaporation**. Once the water cools in the sky, it forms clouds. The clouds fill with drops of water. As the drops get bigger, they get heavy. When they get too heavy to stay in the cloud, they fall to the ground. Depending on how warm or cold it is, the drops may fall as rain or snow.

Hail Cannons

Since the late nineteenth century, hail cannons have been used during thunderstorms to try to protect crops and food from damage. A hail cannon is tall and round. It is said to work by sending a loud blast, called a shock wave, into the air during a thunderstorm. The noise from the blast is believed to break up any

This illustration shows farmers using hail cannons.

hail in the clouds above. The hail turns into rain or tiny bits of ice. Although scientists aren't so sure hail cannons work, many farmers use hail cannons to protect their crops. Some carmakers also use them to protect their new cars from damage during thunderstorms.

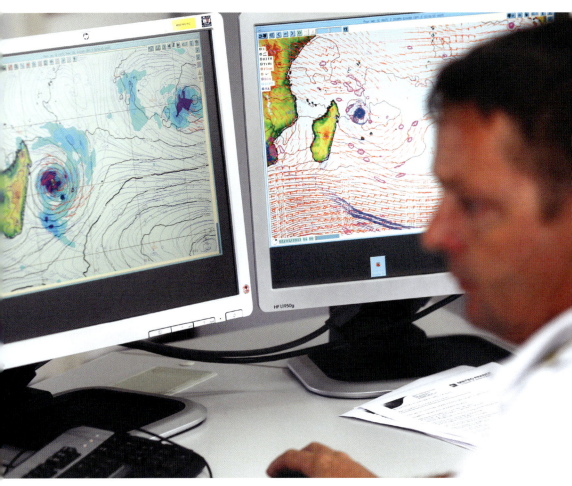

This man tracks weather systems using weather radar.

CONTROLLING THE WEATHER

In order to control the weather, people would need to invent machines and technology that can control the science behind the weather. We would need to be able to fill clouds with drops of water when we want it to rain. If we want snow, we would need to be able to make it cold enough to snow.

Scientists have developed ways to see what kind of weather we will have and when storms are coming. They have invented weather radar devices. Weather radar allows us to see storms on a map. This lets weathermen and women warn people in the path of dangerous storms to take shelter. Seeing what weather is coming is different from controlling the weather. But it is a step in the right direction. Learning about how storms form is an important part of learning to control the weather.

CHAPTER 3

ANIMALS AND INVENTIONS

We know that humans cannot hold electricity without harm. However, there are some animals with electric abilities. This means their bodies have the power to make electricity.

ELECTRIC ANIMALS

Many animals with electric abilities live in the ocean. One animal is the electric eel. Electric

Opposite: Electric eels create electric shocks to hunt for food and protect themselves from creatures that want to eat them.

eels have special parts of their bodies that send out electric shocks. An electric shock is what happens when a large amount of electricity is sent at one time. A person or animal who gets an electric shock becomes still or may even die. Electric eels use the shocks mostly for food and protection. They use their abilities to shock **prey** and **predators**. Prey are animals the electric eel hunts for food. Predators are animals that hunt the electric eel.

Another animal with electric abilities is the electric ray. The electric ray also has special parts of its body that can send out electric shocks. Electric rays use electric shocks for protection. They shock predators to keep them from attacking. It is possible that electric rays may also use their electric abilities to find other rays. They may use electricity as a way to communicate with each other.

DID YOU KNOW?

Small lizards called geckos can climb walls because static electricity helps their feet stick to the wall.

Electric rays live in the ocean and use electric shocks to protect themselves.

PRODUCING ELECTRICITY

Many inventions have been created in the hopes of controlling electricity and the weather. One invention is the **wind turbine**. A wind turbine is a tall tower with large blades attached. It looks like a giant fan. The wind causes the blades to spin around. When the blades spin around, they power a motor. The motor makes electricity. In the past, electricity has been used to power fans to make wind. But with wind turbines, the wind is used to make electricity. Using a natural resource like wind to create something as useful as electricity is a step toward controlling the weather.

This is a field filled with wind turbines. Wind turbines use the wind to create electricity.

Another invention that uses weather to produce electricity is a **solar panel**. "Solar" means coming from the sun. A panel is a large square. A solar panel is a large square that is placed on a roof or an area that gets a lot of sun. The solar panel receives sunlight. Then it turns the sunlight into electricity. Many homes around the world use solar panels as a way to help power their homes.

Wind turbines and solar panels are good for people and for the planet. Wind turbines and solar panels do not create air **pollution** the way some large power plants do. Pollution is when garbage or harmful things are put into the ground or the air. It can lead to damaging the planet and human health. The wind and sun are natural sources of energy. This means humans do not make them, so they do not have to worry about running out of them. These new technologies are also creating

Many people use solar panels on their roofs to help power their homes.

many jobs across the country. Wind turbines and solar panels do not control the weather. But they do use the weather in a positive way.

CREATING LIGHTNING

Scientists have learned to create a form of lightning through special static electricity globes. They are clear glass balls with gases inside. An electric current flows through the gases. When you touch your fingers to the outside of the ball, electrons travel to your fingers. When this

Lightning Rods

Benjamin Franklin used his knowledge of lightning and electricity to make many inventions. After proving that lightning was electrical, he invented the lightning rod. A lightning rod is a large pole. It is usually attached to the roof of a building. It is dangerous for lightning to strike a building. A lightning rod works by carrying a lightning strike away from a building and toward the ground instead. This means less damage to buildings and less danger for people.

This young woman is having fun creating lightning by pressing her hand to an electricity ball.

happens, tiny lightning bolts will appear on the inside of the ball. These are like static electric shocks. These static electricity balls are often found in science museums.

CONTROLLING THE RAIN

For many years, scientists have tried to find a way to control the rain. One way is through something called cloud **seeding**. Seeding means planting something and often refers to planting seeds in the ground. Cloud seeding means planting materials like special ice inside clouds. When the special materials are planted inside the clouds, it can cause the clouds to send rain. Being able to make it rain is one way people are learning to control the weather. If this ability to control the rain continues and grows, we might be able to stop flooding in some areas. We might also be able to end droughts in other areas. A flood happens when an area receives too much water too fast. A drought is what happens when we do not receive enough rain to feed crops, plants, and animals. It can also cause large fires because the ground becomes very dry. Rain is also needed to fill rivers and lakes that people use for water inside their homes.

Scientists can also use cloud seeding to try to prevent rain on certain days. While we need rain, there are times when we don't want it to rain. This is usually on important days with outdoor parties. One important day was the opening of the Olympics in China in 2008. People did not want it to rain on the ceremony. So before the ceremonies began, the Chinese fired over one thousand rockets into the sky. This was a form of cloud seeding.

They hoped to make any rain in the clouds come out. They wanted it to rain before the ceremonies started. It worked. Some cities got a lot of rain after the rockets were fired. But the rain stopped before it reached the city of Beijing, where the ceremonies were being held. The cloud seeding was a success, and it did not rain on the ceremony. However, cloud seeding doesn't always work. It also costs a lot of money. So while it is sometimes used for important events, it is not used often.

CHAPTER 4

INNOVATIONS OF THE FUTURE

People continue to be interested in finding ways to control electricity and the weather. There are many ways that controlling electricity and the weather are helpful for people. These include growing food, avoiding droughts, and making storms safer. In the future, new ways may be invented to control electricity and the weather. What might these new ways or inventions look like?

Opposite: Wind turbines in the ocean have more wind to make electricity than wind turbines on land.

THE FUTURE OF ELECTRICITY

Wind turbines and solar panels are great sources of electricity from nature. What might these technologies look like in the future? One way that solar power may be used in the future is in solar-powered tents. In 2017, a group of young girls invented a solar-powered tent as a science project. The tent worked like solar panels do. It used light from the sun to give the tent electricity. This meant the tent had lights at night when it was dark outside. It also used light from the sun to clean the tent. When the tent was not being used, it could be rolled up into a backpack. This made the tent easy to carry around. The girls invented this tent to help the homeless citizens in their area. They wanted to create a clean place for them to sleep. This kind of tent could also be useful for traveling and camping. It is possible that in the future we will see more inventions using solar power or other natural

DID YOU KNOW?

There are over fifty thousand wind turbines installed across the United States.

Solar-powered tents could become a great resource for campers and for sheltering the homeless.

resources to create electricity. Using solar power to provide for basic needs like shelter is a great reason to work toward controlling the weather and electricity.

Wind turbines have also become a popular source of electricity. What does the future of wind turbines look like? Several wind turbines grouped together make a wind farm. Wind farms are a powerful source of electricity. The idea of building **offshore** wind farms has become popular.

"Offshore" means in the ocean, away from land. There is a lot more wind in the ocean than there is on land. Offshore wind farms would have a lot more wind to power the turbines. This means they would be able to produce a lot of electricity. The electricity could then be used by people on land. The first offshore wind farm to be completed in the United States was turned on in December 2016. It is in the ocean next to Rhode Island. If it is successful at producing more electricity than wind turbines on land do, we might see more offshore wind farms being built in the ocean surrounding the United States.

THE FUTURE OF WEATHER CONTROL

Another reason offshore wind farms might be a good idea is that they may be used to control hurricanes. Hurricanes are very dangerous storms that form over the ocean. When they hit land, they cause a lot of damage. But wind slows down when it goes through the blades of a wind turbine, so it is possible that wind turbines may be able to slow down the dangerous winds of hurricanes. Hurricanes become weaker as their winds slow down. As the hurricanes weaken, there will likely

In the future, lasers could be used to control lightning.

Controlling Lightning

Scientists are working on new ways to create and control lightning. One of these ways is by using lasers. Scientists have learned that when they point special lasers into clouds, they can cause lightning inside the cloud. With this new technology, lasers might be used in the future as a new form of lightning rod. Using lasers, scientists may find a way to point lightning where they want it to go. This would make thunderstorms and lightning much safer for people. Scientists would be able to use lasers to point the lightning away from people and buildings.

Islands in the Caribbean are hit by hurricanes every year. In the future, hurricane winds may be slowed down by wind turbines.

The Science of Controlling Electricity and Weather

be less damage when the storms hit land. Weaker hurricanes would also make it safer for people on land and in the path of the storms. The offshore wind farms would need to be built in places that get a lot of strong winds and hurricanes. If they work, they will be a new way of controlling the weather.

Wind turbines may not be the only way to control weather in the future. Cloud seeding has been a successful way to control the rain. What might it look like in the future? Cloud seeding may be made safer for people in the future by using **drones**. A drone is an aircraft that flies without a human pilot. Cloud seeding can be dangerous. The pilots often have to fly close to mountains and through a lot of wind. These conditions make it hard to fly and put the pilot in danger. If cloud seeding could be done using drones instead of pilots, it would become safer for people. It is possible we will see cloud-seeding drones in the future.

Another way cloud seeding may change in the future is by not using planes or drones at all. There is a form of cloud seeding that uses a laser instead of an aircraft. A laser is a stream of light. Cloud seeding uses an aircraft to release certain materials into the clouds. Some of those materials may be dangerous or cause damage

This drone flies high above the earth. Drones might be used for cloud seeding in the future, making it a safer process for people.

to the environment over time. Lasers have no aircrafts and no materials. They can create clouds and cause water drops to form, just like aircrafts or drones can do. But lasers do this using only light. The light takes electrons out of the air. Then clouds form and water drops form inside the clouds. While drones are safer than pilots for cloud seeding, lasers may be even safer. We may see laser cloud seeding becoming more popular in the future.

GLOSSARY

ATOM A very tiny unit made up of protons, electrons, and neutrons.

BATTERY A container that holds a flow of electricity inside it.

CURRENT Something flowing from one place to another.

DRONE An aircraft that flies without a human pilot.

ENERGY The ability to move around or to move an object.

EVAPORATION When water becomes warm and rises into the sky.

HAIL Frozen rain that falls during a thunderstorm and sometimes before a tornado.

MOTOR A moving tool that creates power for a machine.

OFFSHORE In the ocean; away from land.

POLLUTION Garbage or harmful things in the ground or air that can damage the planet and health of life on Earth.

PRECIPITATION When water falls from clouds in the form of rain, snow, or hail.

PREDATOR An animal that hunts smaller animals.

PREY Animals that are hunted by larger animals.

SEEDING Planting something, usually in the ground.

SOLAR PANEL A large square that takes in a lot of sun and turns it into electricity.

WIND TURBINE A tall tower with blades that create electricity when moved by the wind.

FIND OUT MORE

BOOKS

Demuth, Patricia Brennan. *Thomas Edison and His Bright Idea.* New York: Penguin Young Readers, 2016.

Graham, Ian. *You Wouldn't Want to Live Without Electricity.* New York: Scholastic Children's Press, 2014.

Harris, Caroline. *Discover Science: Weather.* New York: Kingfisher, 2017.

WEBSITES

Easy Science for Kids: Electricity

http://easyscienceforkids.com/electricity

This website explains how electricity works. It also has fun facts about electricity.

History: Nikola Tesla

https://www.history.com/topics/inventions/nikola-tesla

Explore the life and inventions of Nikola Tesla on this website.

Mocomi (Motion Comics for Kids): Hydroelectricity

http://mocomi.com/hydroelectricity

This website includes an animated illustration about what hydroelectricity is and how it works.